A Student's Guide to Natural Science

Stephen M. Barr

D1056119

ISI Books

Wilmington, Delaware

A Student's Guide to Natural Science is made possible by grants from the Lee and Ramona Bass Foundation, the Lillian S. Wells Foundation, and the Wilbur Foundation. The Intercollegiate Studies Institute gratefully acknowledges their support.

Barr, Stephen M., 1953–

    A student's guide to natural science / Stephen M. Barr.—1st ed.—Wilmington, DE : ISI Books, 2006.

        p. ; cm.
        (ISI guides to the major disciplines)

        ISBN-13: 978-1-932236-92-7
        ISBN-10: 1-932236-92-9
        Includes bibliographical references.

        1. Science. 2. Science—History. 3. Physics. 4. Astronomy.
    I. Title.

Q158.5 .B37 2006   2006925553
500—dc22   0608

ISI Books
Intercollegiate Studies Institute
3901 Centerville Road
Wilmington, DE 19807-1938
www.isibooks.org

Design by Sam Torode

CONTENTS

INTRODUCTION

᠅

THE "NATURAL SCIENCES" include physics, astronomy, geology, chemistry, and biology, and are usually referred to as such in contradistinction to the "human sciences," such as anthropology, sociology, and linguistics. Of course, there is some overlap. Those disciplines which study human beings as biological organisms belong to both the natural and the human sciences.

In such a guide as this it would be impossible to give equal attention to all branches of natural science; I have therefore chosen to emphasize physics and, to a lesser extent, astronomy. There are several reasons for this choice. First, breakthroughs in these fields produced the Scientific Revolution and inaugurated the era of modern science. Second, physics can be regarded as the most fundamental branch of natural science, since the laws of physics govern the processes studied in all the other branches. Natural scientists tend to look at things from a "bottom up" perspective, in which the behavior of complex systems is accounted for in terms of the interactions of their constituents, and the branch of science that studies the most basic constituents of matter and their interactions is physics. Third, it can be said that developments in physics and astronomy have had the most profound impact on philosophical thought—along with Darwin's theory of evolution. Finally, there is the fact that I am myself a physicist.

Science was done in each of the great ancient civilizations of Asia, Africa, and the Americas. However, the story of science, as usually told, traces a path from the ancient Greeks and their precursors in Babylon and Egypt, through the Islamic world, and into Europe. There is good reason for this. All of modern science stems from the Scientific Revolution, which erupted in Europe in the 1600s and had its roots in the achievements of the ancient Greeks. The scientific developments that took place in other parts of the world in ancient times, though quite impressive in their own right, made little or no contribution to the Western Scientific Revolution and thus had hardly any lasting impact. (There are exceptions; for example, the concept of the number zero was first developed in India; it made its way into Europe through the Arabs.) From the sixteenth century through the nineteenth, advances in science came almost exclusively from within Europe's borders. It was not until the twentieth century that science became a truly global enterprise.

THE BIRTH OF SCIENCE

᪥

NATURAL SCIENCE IN THE WEST was born in Greece approximately five centuries before the birth of Christ. It was conceived by the coming together of two great ideas. The first was that reason could be systematically employed to enlarge our understanding of reality. In this regard, one might say that the Greeks invented "theory." For instance, while literature is as old as writing, and politics as old as man, political theory and literary theory began with the Greeks. So too did the study of logic and the axiomatic development of mathematics. One of the earliest Greek philosophers, Heraclitus (540–480 B.C.), taught that the world was in constant flux, but that underlying all change is Reason, or *Logos.*

The second great idea was that events in the physical world can be given natural—as opposed to supernatural, or exclusively divine—explanations. The pioneer of this approach was Thales of Miletus (625–546 B.C.), who is said to have explained earthquakes by positing that the earth floated on water. He is most famous for speculating that water is the fundamental principle from which all things come. Thales was thus perhaps the first thinker to seek for the basic elements (or in his case, element) out of which everything is made. Others proposed different elements, and eventually the list grew to four: fire, water, earth, and air.

Stephen M. Barr

The search for the truly fundamental or "elementary" constituents of the world has continued to this day. In 1869, Mendeleev published his periodic table of chemical elements (which at that point numbered sixty-three). Later, the atoms identified by chemists were found to be composed of subatomic particles, which are now studied in the branch of science known as elementary particle physics. Today it is suspected that these particles are not truly elementary but are themselves manifestations of "super-strings." If this present speculation proves to be correct, it will vindicate Thales' intuition that there is but a single truly fundamental "stuff" of nature. In fact, as we shall see, this dream of theoretical unification and simplification has been progressively realized with each of the great advances of modern science.

The idea of "atoms" was the most remarkable and prescient of all the ancient Greek scientific ideas. It was proposed first by Leucippus (fifth century B.C.) and Democritus of Abdera (c. 460–370 B.C.). The Nobel laureate Richard Feynman, in his great *Lectures on Physics*, wrote,

If, in some cataclysm, all of scientific knowledge were to be destroyed, and only one sentence were to be passed on to the next generation of creatures, what statement would contain the most information in the fewest words? I believe it is the atomic hypothesis . . . that all things are made of atoms—little particles that move around in perpetual motion, attracting each other when they are a little distance apart, but repelling upon being squeezed into one another.[1]

4

Of course, the rudimentary version of atomism proposed by

───────────────

ARCHIMEDES (c. 287–212 B.C.), one of the great figures of mathematical history, was born in Syracuse, Sicily. He discovered ways of computing the areas and volumes of curved figures, methods that were further developed by Torricelli, Cavalieri, Newton, and Leibniz in the seventeenth century in order to create the field of integral calculus. Unlike most Greek mathematicians of antiquity, Archimedes was deeply interested in physical problems. He was the first to understand the concept of "center of gravity." He also founded the field of hydrostatics, discovering that a floating body will displace its own weight of fluid, while a submerged body will displace its own volume. He used the latter principle to solve a problem given to him by King Hiero of Syracuse, namely, to determine (without melting it down) whether a certain crown was made from pure or adulterated gold. Hitting upon the solution while in the public baths, he ran naked through the streets shouting "Eureka!" ("I have found it!"), the eternal cry of the scientific discoverer.

Archimedes was also the discoverer of the principle of the lever, boasting, "Give me a place to stand and I will move the earth." Legend has it that he helped defend Syracuse from a Roman siege during the Second Punic War by inventing fantastic and ingenious weapons, such as the "claw of Archimedes," and huge focusing mirrors to ignite ships. According to Plutarch, "[Archimedes] being perpetually charmed by his familiar siren, that is, by his geometry, neglected to eat and drink and took no care of his person;...[he] was often carried by force to the baths, and when there would trace geometrical figures in the ashes of the fire, and with his fingers draw lines upon his body when it was anointed with oil, being in a state of great ecstasy and divinely possessed by his science." During the siege of Syracuse, in spite of standing orders from the Roman general that the great geometer not be harmed, Archimedes was struck down by a Roman soldier while drawing geometrical diagrams in the sand. His last words were, "Don't disturb my circles."

Leucippus and Democritus was not a scientific theory in our modern sense. It could not be tested, and it led to no research program, but rather remained, as did most of Greek natural science, at the level of philosophical speculation.

The beginning stage of any branch of science involves simple observation and classification. Not surprisingly, much of Greek natural science consisted of this kind of activity. At times it was more ambitious and sought for causes and principles, but these principles were for the most part philosophical. In other words, they were not formulated into scientific laws in the modern sense. One thinks of Aristotle's principle that "nothing moves unless it is moved by another." This was meant as a general statement about cause and effect. It did not allow one to predict anything, let alone to make calculations.

It is interesting that the Greeks, for all their tremendous achievements in mathematics, did not go far in applying mathematics to their study of the physical world—astronomy being the major exception. This should not surprise us. It is perhaps obvious that the world is an orderly place, as opposed to being mere chaos; but the fact that its orderliness is mathematical is very far from being obvious, at least if one looks at things and events on the earth, where there is a great deal of irregularity and haphazardness. The first person to conceive the idea that mathematics is fundamental to understanding physical reality—rather than pertaining only to some ideal realm—was Pythagoras (c. 569–c. 475 B.C.). This insight was perhaps suggested to him by his research in music, where he discovered that harmonious tones are produced by strings whose lengths are in simple arithmetical ratios to each other. In any event, Pythagoras and his followers arrived at the idea that reality at its deepest level is mathematical. Indeed, Aristotle attributed to the Pythagoreans the idea that "things are numbers." This assertion may seem extreme, and doubtless did to Aristotle, but to the modern physicist it appears both profound and prophetic.

It is in the motions of the heavenly bodies that the mathematical orderliness of the universe is most apparent. This has to do with a number of circumstances. First, interplanetary space is nearly a vacuum, which means that the movements of the solar system's various bodies are unimpeded by friction. Second, the mutual gravitation of the planets is small compared to their attraction to the sun, a fact that greatly simplifies their motion. In other words, in the solar system nature has provided us with a dynamical system that is relatively simple to analyze. This was vital for the emergence of science. In empirical science it is important to be able to isolate specific causes and effects so that they are not obscured or disrupted by extraneous and irrelevant factors. This normally has to be done by conducting "controlled" experiments (experiments, for instance, that allow one to compare two systems that differ in only one respect). Usually, it is only in this way that one gets a chance to observe interesting and significant patterns in the data. But it did not occur to the ancient Greeks to perform controlled experiments—or, for the most part, experiments of any kind. It is therefore very fortunate that they had the solar system to observe.

The first application of geometry to astronomy seems to have been inspired by Pythagorean ideals. Pythagoras himself suggested that the earth is a perfect sphere. Later, Eudoxus (c. 408–c. 355 B.C.) proposed a model in which the apparently complex movements of the heavenly bodies resulted from their motion in perfectly circular paths. These Pythagorean principles—that theories should be mathematically beautiful and that they should explain complex effects in terms of simple causes—have been tremendously fruitful in the history of science. But they are not sufficient. The mathematical approach of these earlier astronomers lacked a key ingredient, namely the making of precise measurements and the basing of one's theories upon those measurements. In this respect Hipparchus (c. 190–c. 120 B.C.) far exceeded his predecessors and transformed astronomy into a quantitative and predictive science. He made remarkably accurate determinations of such quantities as the distance of the moon from the earth and the rate at which the earth's axis precesses (the so-called precession of the equinox, a phenomenon that he discovered). After Hipparchus, the development of ancient Greek astronomy reached its culmination in the work of Ptolemy (c. 85–c. 165), whose intricate geocentric model of the solar system was to be generally accepted for the next fifteen centuries.

The Greeks' attraction to mathematics was a double-edged sword. On the one hand, it had incalculable benefits for science. The Greeks' most enduring scientific legacy lay in their mathematics and mathematical astronomy. On the other hand, this attraction to mathematics reflected a tendency (seen very markedly in Plato) to disdain the world of phenomena for a more exclusive focus on the realm of the ideal.

We find quite the opposite tendency in Aristotle. Aristotle was very much interested in phenomena of every kind, and (in contrast to many of his later epigones) engaged in extensive empirical investigations, especially in biology.

HIPPARCHUS (c. 190–120 B.C.) is considered the greatest observational astronomer of antiquity. Little is known of his life except that he was born in Nicaea, located in present-day Turkey, and spent most of his life on the island of Rhodes. What distinguished him from his predecessors was his application of precise measurements to geometrical models of astronomy. Not only did he make extensive measurements; he also made use of the voluminous astronomical records of the Babylonians, which dated back to the eighth century B.C. This long span of data allowed him to compute certain quantities with unprecedented accuracy.

Hipparchus created the first trigonometric tables, which greatly facilitated astronomical calculations, and developed or improved devices for astronomical observation. He also compiled the first star catalogue, which gave the positions of about one thousand stars. Though he worked on many problems, such as determining the distance to the moon, he is most famous for discovering the "precession of the equinox" and correctly attributing it to a wobbling of the earth's axis of rotation. Newton later showed that this wobbling was caused by the gravitational torque exerted by the sun and moon on the earth's equatorial bulge.

He was undoubtedly one of the greatest biologists of the ancient world. His legacy in physics, however, is more ambiguous—indeed, on the whole, perhaps negative. There are several reasons for his comparative failure as a physicist. First, there is the fact, already remarked on, that terrestrial phenomena are hard to sort out. Among many other complications, they involve large frictional forces, which resulted in Aristotle being fundamentally misled about the relationship between force and motion. Second, Aristotle appreciated neither the true nature of mathematics nor its profound importance; his genius lay elsewhere. Third, Aristotle's approach to the physical sciences was philosophical; in his thought there is no bright line between physical and metaphysical concepts. This would not have created so many problems for later thought—problems discussed in more detail below—had it not been for the very brilliance and depth of Aristotelian philosophy.

THE SECOND BIRTH OF SCIENCE

IT IS SOMETIMES SAID, by those with an axe to grind against religion, that the rise of Christianity brought an end to the first great age of scientific progress. This claim is untenable. It is true that one can find statements in the writings of the church fathers that deprecate the study of nature, and that science was not high on the early Christians' list of concerns. However, one finds the same range of attitudes toward science among the early Christians as among their pagan contemporaries. And the fact is that the glory days of ancient science were long gone by the time Christians became a significant demographic or intellectual force. The golden age of Greek mathematics ended two hundred years before the birth of Christ. (For example, the great Greek mathematicians Archimedes, Eratosthenes, and Apollonius of Perga died, respectively, in 212 B.C., 194 B.C., and 190 B.C..) Only a few great figures in ancient Greek science date from the period after Christ, notably the astronomer Ptolemy, who died around A.D. 165, and the mathematician Diophantus, who died around A.D. 284. At that point, Christianity was still a small and persecuted sect.

As is well known, an impressive revival of mathematics and science began in the Islamic world in the ninth century. Under the Abbasid caliphate, which stretched from North Africa to Central Asia, scholars were able to draw upon the patrimony of the Babylonians and Indians as well as the Greeks. The Muslim contributions to science are memorialized in the many scientific terms of Arabic origin, such as *alcohol* and *alkali* in chemistry (a field of inquiry once called "alchemy"); *algebra, algorithm,* and *zero* in mathematics; and *almanac, azimuth, zenith,* and the names of the bright stars *Algol, Aldebaran, Betelgeuse, Rigel,* and *Vega* in astronomy. However, the brilliance of Muslim science began to fade after a few centuries. The Islamic theological

establishment tended to be indifferent or hostile to speculative Greek thought, and therefore science did not achieve
the kind of institutional status in the Muslim world that
it later achieved in the universities of Europe.

The second birth of science really came in the Latin
West. In the eleventh century, when Western Europe began
to recover from the economic and cultural collapse caused
by the barbarian invasions, its scholars became aware again,
largely through contact with the Arab world, of the ancient
Greeks' great achievements in science. This awareness
engendered an insatiable curiosity about and demand for
the works of ancient Greek scholars, which led in turn to a
frenzy of translations of these works into Latin, either from
Arabic sources obtained in Spain or directly from Greek
versions obtained from the Byzantines. Universities were
invented in medieval Europe, and they were founded in
part as places where this newly recovered knowledge could
be studied. The intense interest in Greek science—or, as it
was called at that time, "natural philosophy"—was shared
by clergy and laity alike. Indeed, in medieval universities
the study of natural philosophy was a prerequisite for the
study of theology. (This would be somewhat analogous to
physics being a required course in today's seminaries.)

For a long time, it was standard for modern scholars
to dismiss medieval science as lacking in creativity or true
scientific spirit, and as being quite irrelevant to later scientific progress. However, scholars such as Pierre Duhem
and A. C. Crombie successfully challenged that consensus.

They demonstrated that medieval science was far more vital than had been supposed, and that the picture of monkish scholars slavishly following Aristotle had been overdrawn. The "natural philosophers" of the Middle Ages were quite aware of some of the inadequacies in Aristotle's ideas and adopted a cautiously critical approach to him, though their interesting critiques were not based on experiments but on logical reasoning—and to some extent on what we would today call "thought experiments." In addition, the medievals took tentative steps toward developing a science of motion. The crucial concept of uniform acceleration (or in their quaint terminology, "uniformly difform" motion) became understood for the first time; the notion of "impetus" (an anticipation of the concept of "momentum") was developed; graphs were invented to facilitate reasoning about mathematical functions and motion; and mathematical laws of motion were first proposed. Some historians, such as Duhem, Crombie, and more recently Stanley Jaki, have even claimed that these ideas directly influenced the thinking of Galileo and other founders of the Scientific Revolution (though the extent of this influence is disputed, and the issue is far from settled).

Be that as it may, there is one way in which the revival of science in medieval Europe certainly did lay the groundwork for the Scientific Revolution to come. It "institutionalized" science, as Edward Grant, the noted historian of science, has put it. In the ancient and Arab worlds, science like art had depended upon the patronage of wealthy or powerful individuals who happened to have a personal interest in it. It was therefore a hit-or-miss affair, subject to the vicissitudes of politics and economic fortunes. By contrast, in the medieval universities there was created for the first time a stable community of scholars that studied scientific questions continuously from generation to generation. That is, a scientific community came into being. By the end of the Middle Ages there were nearly one hundred universities in Europe, and their graduates numbered in the tens of thousands. This created a significant literate public that was interested in science, was willing to pay to be taught or obtain books about it, and from whose ranks scientific talent could emerge.

Without the scientific community and the scientific public created by the medieval universities, the Scientific Revolution would not have had fertile soil in which to germinate.

SCIENCE, RELIGION, AND ARISTOTLE

THE FOREGOING DISCUSSION raises two very interesting and difficult questions: Why did the Scientific Revolution occur in Europe, and what was the role of religion in that revolution? One view, encountered so frequently that it has become a cliché, is that the Christian religion was the enemy of science and tried to strangle it at its birth; this animus is alleged to have been clearly revealed in the Galileo affair. However, scholars no longer take this view seriously. The idea that the church establishment has been implacably hostile to science is a myth created to serve the purposes of antireligious and anticlerical propaganda.

In fact, the church has always esteemed scientific research—even at the time of the Galileo affair. We have seen that the medieval church was willing to embrace the science of the ancient Greeks, even though it was naturalistic in character and pagan in origin, and that this science had an important place in the curricula of medieval universities, institutions that had been founded primarily under church auspices, received much church patronage and protection, and were staffed largely by clerics. Indeed, most of the scientists of the Middle Ages were clergymen, such as Nicholas Oresme (c. 1323–82), who was bishop of Lisieux and a mathematician and physicist of great ability. This tradition of clergy involvement in scientific research has continued to the present day. In fact, from the seventeenth through the twentieth century, a remarkable number of important scientific contributions have been made by Catholic priests.[2]

The favorable attitude of the church toward "natural philosophy" was largely the result of the efforts of St. Albert the Great, who helped introduce Greek science into the medieval universities, and his pupil St. Thomas Aquinas. Both men were convinced of the possibility and importance of harmonizing faith and reason and saw in the philosophy of Aristotle the conceptual tools needed to accomplish that. This gave a tremendous impetus to the study of natural phi-

---

ORESME, NICHOLAS (c. 1323–82) was born near Caen, Normandy, received his doctorate in theology from the University of Paris, became counselor and chaplain to the king of France, and was ultimately installed as the bishop of Lisieux. A polymathic genius, he made pioneering contributions in mathematics, physics, musicology, and psychology. He was an important figure in scholastic philosophy, and is considered by many to be the greatest economist of the Middle Ages. In mathematics, he was the first to discuss fractional and even irrational exponents, and his studies of infinite series proved that the harmonic series $(1 + 1/2 + 1/3 + 1/4 + \ldots)$ diverges. Oresme was also the first to use graphs to plot one quantity as a function of another, extending his discussion to three-dimensional graphs, thus anticipating by three centuries some of the key ideas of Cartesian analytic geometry. He used such graphical methods to give the first proof of the "Merton theorem," which provides the distance traversed by a uniformly accelerated body. These studies probably contributed indirectly to Galileo's discovery of the law of falling bodies.

In addition to all this, Oresme proved that phenomena could be accounted for satisfactorily by assuming that the earth rotates rather than the heavens. The analysis by which he refuted common physical objections to this was superior to those later articulated by Copernicus and Galileo, especially in its resolution of motion into vertical and horizontal components. He was the first to understand the distortion of the apparent positions of celestial objects near the horizon as resulting from the bending of light passing through air of varying density. And he argued for the existence of an infinite universe, contrary to the standard Aristotelian view. All in all, he must be accounted one of the great original thinkers in the history of mathematics and physics—and an important forerunner of the Scientific Revolution.

losophy and thereby helped prepare the way for the Scientific Revolution. However, the church's embrace of Aristotle also had negative consequences, since it helped to cement in place a mistaken approach to the physical sciences.

The philosophical system of Aristotle, as transformed and Christianized by St. Thomas Aquinas, was a brilliant and impressive intellectual synthesis. One can think of it as a medieval "grand unified theory" or "theory of everything," except that it comprised not only natural philosophy, but also metaphysics, anthropology, moral philosophy, and much else, including even truths of revealed religion (insofar as they could be grasped by human reason). This Aristotelian-Thomistic philosophy has great strengths and remains indeed very much a living tradition. However, for all the light Aristotle's ideas shed upon metaphysical and moral issues, they contained much that was deeply mistaken and misleading when it came to physics and astronomy.

It is foolish to fault the medievals for adopting Aristotle's physics, for there was nothing else around; it was the science that they inherited from the Greeks. And for a time it did play the useful function (as even wrong theories can) of providing a framework for theoretical discussion and analysis. It also provided an example of a naturalistic theory of physical phenomena, which is no small thing. Nevertheless, it had the effect of leading scientific thought into a cul-de-sac from which it took considerable effort to escape. To the extent that theology contributed to prolonging the dominance of Aristotelian natural philosophy, it played an unhelpful role.

The dominance of Aristotelianism helps to explain the church's condemnation of Galileo and heliocentric astronomy in 1633. However, it was only one factor in a very complex affair. The other reasons for Galileo's con-

---

GALILEI, GALILEO (1564–1642) was born in Pisa and began studies at the University of Pisa in 1581. He secured a professorship there in 1589, but decided to move to the University of Padua two years later because of conflicts with Aristotelians. In 1609, having heard of the invention of the telescope, he devised his own and with it began studying the heavens. His discoveries of sunspots, the moons of Jupiter, mountains on the moon, and the phases of Venus undermined Aristotelian science, refuted the Ptolemaic system, and made him a celebrity. He aroused opposition by his advocacy of the heliocentric Copernican system, and in 1616 the Roman Inquisition issued an injunction forbidding him to defend Copernicanism "in any way." At the same time, all books were prohibited that advocated Copernicanism as true rather than merely as a "hypothesis" (by which was meant a mathematical device for simplifying calculations). In 1623, Maffeo Cardinal Barberini, a friend and protector of Galileo, was elected pope. Unaware of the 1616 injunction, he did not object to Galileo defending Copernicanism as a "hypothesis." Galileo proceeded to publish in 1630 his *Dialogue Concerning the Two Chief World Systems,* in which he not only defended Copernicanism as true, but seemed also to lampoon the pope's philosophical opinions. The pope was outraged at this betrayal by someone he had protected; the forgotten 1616 injunction was discovered in the files; and in 1633 Galileo was forced to publicly renounce Copernicanism and sentenced to lifelong house arrest, which he served in his villa in Florence, where he was allowed to receive visitors and publish on other scientific subjects. In 1638 he published his *Dialogues Concerning Two New Sciences,* which set out his discoveries in physics, his greatest contributions to science.

demnation included professional rivalry, Galileo's talent for making enemies, and, most important of all, the turbulence of the times. It was an era of great religious tension; Europe was being torn apart by the Thirty Years War, which had begun as a Catholic-Protestant struggle. As part of its effort to defend itself against the Protestant challenge, the Catholic Church had enacted at the Council of Trent (1545–63) a set of rules for the interpretation of scripture that was intended to prevent radical theological innovations. Though reasonable in themselves, these rules ended up being misapplied to Galileo, who had unwisely allowed himself to be drawn into scriptural and theological debate by his enemies.

The condemnation of Galileo, though a fateful blunder, was not the result of hostility toward science on the part of church authorities, nor did it reflect an unyielding dogmatism in scientific matters. The words of Cardinal Bellarmine, head of the Roman Inquisition at the time of Galileo's first encounter with it in 1616, are well worth remembering:

> If it were demonstrated that [the sun was really motionless and the earth was in motion] we should have to proceed with caution in interpreting passages of Scripture that appear to teach the contrary, and rather admit that we do not understand them than declare something false which has been proven to be true.

Bellarmine went on to say that he had "grave doubts" that such a proof existed and that "in case of doubt" one must stay with the traditional interpretation of scripture. As a matter of fact, such a proof of heliocentrism did not exist in Galileo's time, though there were strong indications in its favor for those with eyes to see them.

In any event, if we look at the eight-hundred-year record of the church's involvement with science, it is hard to see the Galileo affair as anything but an aberration. Far from seeing religion as an obstacle to the emergence of modern science, some scholars have argued that Christian beliefs played a part in making the Scientific Revolution possible. A great deal can be said for this view. For example, an idea fundamental to all science is that there exists a natural order. That is, not only is the world orderly, but it is in fact a *natural* world. We have seen that this idea arose among the pagan Greek philosophers, but Judaism and Christianity also helped promote this way of thinking. Whereas in primitive pagan religion the world was imbued with supernatural and occult forces and populated by myriad deities—gods of war, gods of the ocean and of the earth, goddesses of sex and fertility, and so forth—Jews and Christians taught that there was only one God who was to be sought not *within* nature and its phenomena and forces, but outside of nature, a God who was indeed the author of nature. In this way, biblical religion desacralized and depersonalized the world. To borrow Max Weber's term, it "disenchanted" the world.

For example, the book of Genesis, which is often seen as an instance of primitive mythmaking, was actually written in part, scholars tell us, as a polemic against pagan supernaturalism and superstition. When Genesis says that the sun and moon are merely "lamps" placed by God in the heavens to light the day and night, it is attacking the pagan religions that worshipped the sun and moon. When it says that man is made "in the image of God" and is to exercise "dominion" over the animals, Genesis is, among other things, attacking the paganism in which men worshipped and bowed down to animals or to gods made in the image of animals.

Medieval Christians were so comfortable with a naturalistic view of the physical world that it was commonplace as early as the twelfth century, according to Edward Grant, for philosophers and theologians to refer to the universe as a *machina*, a "machine." (Of course, both Judaism and Christianity teach the possibility of miracles. But miracles—from the Latin *mirari*, "to wonder at"—are such precisely because they are rare occurrences that contravene what the medieval philosophers called the "course of nature." They derive their whole significance from the fact that there is a natural order that only God, as author of nature, can override.)

A second idea that Judaism and Christianity likely helped to foster is that the universe is not merely orderly but *lawful*. A Christian writer of the second century, Minucius Felix, wrote:

If upon entering some home you saw that everything there was well-tended, neat, and decorative, you would believe that some master was in charge of it, and that he was himself much superior to those good things. So too in the home of this world, when you see providence, order, and law in the heavens and on earth, believe that there is a Lord and Author

---

STENSEN, NIELS (1638–86, also called Steno) made fundamental contributions to anatomy, geology, paleontology, and crystallography. Born in Denmark, he enrolled in the University of Copenhagen to study medicine. In his twenties he was already recognized as one of the leading anatomists in Europe. His anatomical studies greatly increased knowledge of the glandular-lymphatic system. (Stensen's duct, Stensen's foramina, and Stensen's gland are all named after him.) He also did important research on heart and muscle structure, brain anatomy, and embryology. After spending time in Paris and the Netherlands, he traveled to Florence to join the Accademia del Cimento, which had been founded by Ferdinando d'Medici, Grand Duke of Tuscany, to carry out experimental research in the tradition of Galileo, some of whose students were among its members. While dissecting the head of a great white shark that had been caught off Livorno, Stensen realized that the teeth bore a remarkable resemblance to the "tongue stones" found in great abundance in Malta. This led him to undertake those investigations for which he is now regarded as the founder of the scientific study of fossils and of the branch of geology called stratigraphy. His ideas on how geological strata form played a crucial role in unlocking the history of the earth and eventually in revealing the planet's great age. In addition, Stensen discovered the basic law of crystallography, which is that the angles of a given mineral are always the same ("Steno's law"). Stensen converted to Catholicism in 1667, was ordained a priest in 1675, and became a bishop two years later. As a bishop he was given to rigorous asceticism and was an ardent champion of the poor. He was beatified by Pope John Paul II in 1988.

of the universe, more beautiful than the stars themselves and
the various parts of the whole world.

This emphasis on law goes back to the Hebrew scriptures,
the first five books of which are indeed called by Jews the
Torah or "Law." Of course, Israel regarded God as its lawgiver, but he is also thought of as lawgiver to the cosmos
itself. God says in the book of Jeremiah, "When I have no
covenant with day and night, and have given no laws to
heaven and earth, then too will I reject the descendants of
Jacob and of my servant David." Psalm 148 tells of the sun,
moon, stars, and heavens obeying a divinely given "law that
will not pass away." The ancient rabbis said, "The Holy
One, blessed be he, consulted the Torah when he created
the world." The Torah was thus a law that existed eternally
in the mind of God and according to which the universe
itself was made. (This notion of an eternally preexisting law
became linked in Jewish thought to the idea of the divine
Wisdom, personified in later books of the Old Testament,
especially Proverbs, Wisdom, Sirach, and Baruch. This
divine Wisdom became in the New Testament the divine
Logos, which means "Word" or "Reason." In other words,
the idea that reason underlies the universe emerged within
both pagan Greek and Jewish thought.)

It is not only religious authors who have seen the idea of
a divine lawgiver as a factor contributing to the emergence
of modern science. In explaining why Chinese civilization, with all its refinement and splendid achievements,

did not produce a Newton or a Descartes, the decidedly nonreligious biologist E. O. Wilson pointed to the fact that Chinese scholars had

> abandoned the idea of a supreme being with personal and creative properties. No rational Author of Nature existed in their universe; consequently the objects they meticulously described did not follow universal principles.... In the absence of a compelling need for the notion of general laws—thoughts in the mind of God, so to speak—little or no search was made for them.[3]

The cosmologist Andrei Linde, himself an atheist, has also suggested that the idea of a universe "governed by a single law in all its parts" has some connection to monotheism.

On the other hand, of itself monotheism is not enough to produce a scientific revolution. Islam is monotheistic, but the progress of science in Muslim lands eventually petered out. And Byzantine civilization, even though it was Christian and had never forgotten the science of the ancient Greeks, produced little science of its own. It may well be that the especially strong emphasis in the theology of the Western church on law, system, and reason contributed decisively to the emergence of modern science.

But something else was also at work during the Renaissance: a spirit of innovation, restlessness, and questioning that had to see things for itself and was skeptical of received opinion. For some, this even extended to theological matters—although, contrary to what many imagine, religious skepticism does not appear to have been a generative factor in the Scientific Revolution. Almost all of the great founders of modern science, including Copernicus, Kepler, Galileo, Boyle, Hooke, and Newton, were religiously devout. Some of them, like Kepler and Boyle, were clearly inspired to do scientific research by their religious beliefs whereas none of them were motivated by opposition to faith or orthodoxy. Until at least the mid-nineteenth century, most great scientists were religious believers; many continue to be so today.

THE SCIENTIFIC REVOLUTION

THE SCIENTIFIC METHOD

The Scientific Revolution was characterized by three great achievements. First, there was the shattering of the Aristotelian synthesis and a decisive break with its philosophical, speculative, and qualitative approach to doing science (even though science continued to be referred to as "natural philosophy"). Second, there was the realization of the importance of doing experiments and making precise measurements. This involved the growing use of artificial devices such as telescopes, pendulums, and vacuum pumps in investigating nature. (This second achievement also required a conceptual advance; in Aristotelian thought, machines had been regarded as causing things to move contrary to their "natural" motion.) And, finally, there was the mathematization of science. Of course, science had previously employed mathematics and even experiment, but it was in the 1600s that these tools came together to create a powerful new way of investigating the world, often called "the scientific method."

The scientific method comprised (a) the collection of data by precise measurements and controlled, repeatable experiments, (b) the formulation of testable hypotheses to explain either regularities or anomalies in the data, and (c) the verification or falsification of those hypotheses by comparing their predictions with the results of new measurements or experiments. Unfortunately, the scientific method is sometimes spoken of as if it were an automatic or mechanical process. It is not a *process*, but an *activity*. Processes can be undertaken by machines. Activities require the imagination, insight, cleverness, initiative, creativity, and judgment of persons.

It is not sufficiently stressed in scientific education, where the main object is to master an enormous body of fact and theory, just how marvelously ingenious experiments and observations can be. Great experiments, like great theoretical ideas, are things of beauty. However, they tend to be more ephemeral. They are the scaffolding in the construction of the edifice of science and are too often forgotten after they have played their part.

Many also have an unrealistic understanding of scientific theorizing. Students are usually presented with a theory as a finished product, with all obscurities removed, essential ideas crystallized into precise concepts, fundamental principles identified, and logical structure clarified. This is indeed the best way for students to master the ideas. However, it can also lead them to lose sight of the messy, confusing, and arduous struggle by which the key insights were originally won.

A strong case can be made that a lack of appreciation of the actual methods of science has had harmful effects on philosophy. The two great contrary movements in European philosophy in the seventeenth and eighteenth centuries—rationalism and empiricism—were both to some extent inspired by the progress of science and considered themselves scientific in spirit. However, each exaggerated one pole of scientific thought at the expense of the other. The rationalists tended to think of all knowledge as advancing by a process of deductive reasoning from first principles, as in mathematics. They undervalued the empirical component of scientific progress. On the other hand, the empiricists underappreciated the role of hypothesis, abstraction, and theoretical construction in advancing scientific knowledge. Like the rationalists, they may also have been misled by a false analogy with mathematics. In mathematics, each conclusion must be firmly demonstrated before it can be used as the basis for further reasoning. Some have imagined that empirical science works in a similar way. They think that the existence of every entity posited by a theory and the truth of every theoretical proposition must be directly verified before the next step is taken. That is not how things work.

Take, for example, the theory of electromagnetism formulated by James Clerk Maxwell in the mid-nineteenth century. It posits the existence at every point in space and time of a three-component "electric field" and a three-component "magnetic field." No one has ever directly verified the existence of such entities everywhere in space-time, nor would it be possible to do so. However, it is not necessary. That is not how the validity of Maxwell's theory was established in the first place or why physicists still believe it to be true. Maxwell's theory, like any sophisticated scientific theory, is a highly elaborate and abstract structure that presupposes the existence of many things, not all of which can be directly or separately observed. Rather, it is the *theory as a whole* that is verified on the basis of observations; and these observations, however numerous they may be, are necessarily few compared to the entities that the theory assumes to exist.

The empiricists seemed to think that humans built up a picture of reality by adding together a large number of sensory "impressions," from which more complex ideas got generated by a process of "association." However, one does not directly sense magnetic fields (unless one is a monarch butterfly, say). One infers their existence by their often very indirect effects, and even then only with the help of abstract theory. And yet these magnetic fields are as real and as physical as rocks and trees. (The same point is illustrated by the electromagnetic spectrum: we directly sense "visible light" with our eyes, but can only infer the reality of ultraviolet light and radio waves by indirect means. However, this difference is due simply to the characteristics of our sensory organs. Radio waves, ultraviolet light, and visible light are in themselves equally real, differing only in wavelength.) The crude notion of verification that one finds implicitly in the British empiricists of the eighteenth century also afflicted later versions of empiricism, such as the early-twentieth-century philosophical school called "logical positivism." For the logical positivists, every meaningful statement had to be translatable into statements about sensory impressions.

The relationship between theory and observation in science is complex and dynamic. Theories are built on experiment, and experiments depend for their interpretation upon theory. This "theory-dependence of experiment" is much talked about in recent times and has provided an opening for some "postmodern" thinkers to claim that the scientific method involves a "vicious circle" that somehow vitiates the notion of scientific objectivity. One can see through such sophistry by a simple analogy: maps were made by explorers; and explorers had to make use of existing maps. This "circularity" obviously did not prevent better and better maps from being made, nor does the dynamic interaction of theory and experiment prevent better and better theories of the physical world from being made.

Indeed, that is precisely how they must be made; and the recognition of this fact was the fundamental achievement of the Scientific Revolution of the seventeenth century.

Before we turn to the history of that revolution, it is worth saying a bit more about how scientific theories are verified. In mathematics a theorem may be proven rigorously, in which case it can be affirmed with certainty; or it can be disproved, in which case it can be denied with certainty. However, in natural science, as in life, one accepts a theory with a degree of confidence that is, as it were, a continuous variable. In some cases, the confidence may be so great that we can speak of virtual certainty—indeed, of a "scientific fact." For instance, no scientist seriously doubts anymore that matter is made of atoms or that the earth rotates on its axis. In other cases, the confidence of scientists in a theory may be fairly strong, but not so strong that they would say it has been "confirmed." In yet other cases, there exists simply what lawyers would call a "rebuttable presumption" in favor of a theory. It all depends on the quantity and type of evidence.

What counts as evidence for a theory? It is not only a matter of quantitative predictions confirmed by experiment. For one thing, a particular experimental number, or even many such numbers, might be accounted for in several ways. For example, one of the major successes of Einstein's theory of gravity (the so-called general theory of relativity) was that it predicted accurately the "precession of the perihelion of Mercury" (the slow shift over time of the point of closest approach of Mercury to the sun). However, one could have accounted for the perihelion shift in several other ways. One way was to posit a certain amount of solar oblateness (i.e., a flattening at the sun's poles). Another was to posit a certain density of matter filling the space between Mercury and the sun. Nowadays, enough is known about the sun and its environment that these alternative explanations are no longer viable. But it is possible to find other alternatives; one could simply add, for instance, a new term to Newton's inverse-square law of gravity to account for the perihelion shift. So the successful prediction of that shift, though vitally important, was not the only reason that physicists began to believe in Einstein's theory.

Many considerations influence scientists' judgments about the plausibility or likelihood of a theory. These include the theory's *simplicity and economy*, its provision of a *more unified and coherent picture* of nature, its *explanatory power*, its *mathematical beauty*, its grounding in *deep principles*, its prediction of *new phenomena*, and its ability to *resolve theoretical puzzles or contradictions*. Einstein's theory of gravity, for instance, resolved the problem that Newton's theory of gravity was not consistent with the principles of the special theory of relativity. It also explained why "inertial mass" is equal to "gravitational mass." It predicted the phenomenon of the bending of rays of light by gravity. It flowed from the deep "equivalence principle." It was based on the beautiful idea that gravitation is the result of the curvature of space-time. In other words, Einstein's theory was supported by many *converging* lines of evidence and grounds of credibility.

Another good example of how a theory comes to enjoy acceptance is the Dirac equation, invented by P. A. M. Dirac in 1928 to describe electrons in a way consistent with both special relativity and quantum theory. Dirac was led to the equation primarily by considerations of mathematical beauty. But the equation also resolved a puzzle—namely, that the magnetic moment of the electron was twice as big as previous theory had held it ought to be. The Dirac equation predicted a new phenomenon: the existence of anti-particles. And it shed light on a property of electrons called spin. Eventually it was used to make many very precise experimental predictions that were later confirmed. This example illustrates the fact that, while many factors can lead theorists to entertain a certain hypothesis or build their confidence in it, ultimately it is usually a significant number of precise and correct quantitative predictions that "clinches it." (That, at least, is the case in sciences where controlled and repeatable experiments are possible. However, it is unreasonable to demand the same kinds of confirmation in all fields. For instance, there is a great deal of converging evidence that the evolution of species has taken place, but one cannot predict how particular lineages will evolve. Similarly, one may learn what causes earthquakes without being able to forecast them accurately.)

From Copernicus to Newton

In a certain sense, one could almost say that Sir Isaac Newton (1643–1727) *was* the Scientific Revolution. There is much truth in Alexander Pope's famous couplet:

> Nature and Nature's laws lay hid in night.
> God said, "Let Newton be," and all was light.

Newton was a towering peak. There was no one to rival him in physics until the twentieth century. One may think of everything that went before Newton as having set the stage for his great breakthroughs, and everything that came after him—until the twentieth century—as having exploited those breakthroughs.

Three lines of development led to the achievements of Newton: in astronomy, the discovery of Kepler's laws of planetary motion; in physics, Galileo's discovery of the law of falling bodies; and in mathematics, the development of analytic geometry and the use of coordinates by Descartes (1596–1650).

In astronomy, the line that led to Newton began with Copernicus (1473–1543), who sparked the Scientific Revolution with his heliocentric theory of planetary motion. It proceeded through the extremely precise observational work of the great Danish astronomer Tycho Brahe (1546–1601). And it culminated in the discovery by Johannes Kepler (1571–1630) of his three great laws of planetary motion (which would have been impossible without Brahe's data).

Galileo (1564–1642) was not important in this particular line of development—indeed, he firmly rejected Kepler's crucial idea of elliptical planetary orbits. Rather, Galileo's great contribution to astronomy was the use of telescopes, by means of which he made a series of dramatic discoveries—such as the phases of Venus, the moons of Jupiter, and sunspots—that helped undermine Aristotelianism and the Ptolemaic system. In astronomical *theory*, however, it was Copernicus and Kepler, not Galileo, who made the key advances. On the other hand, in physics Galileo made the important breakthrough, when he discovered the law of falling bodies by applying to terrestrial phenomena the powerful combination of experimentation (carried out with inclined planes and pendulums) and mathematics.

COPERNICUS, NICOLAUS (1473–1543) was born in Torun, Poland, and studied astronomy at the University of Cracow. His uncle, the bishop of Ermland, obtained for him a position as canon of the Cathedral of Frauenburg, an administrative office. Copernicus studied civil and canon law at the University of Bologna and medicine at the University of Padua before obtaining a doctorate in canon law from the University of Ferrara in 1503. He thereupon returned to Ermland, where he acted as advisor to his uncle the bishop and took up his duties as canon. He acquired a wide reputation as an astronomer and was visited in 1539 by Georg Joachim Rheticus, professor of mathematics at the University of Wittenberg, who persuaded Copernicus to publish his ideas on heliocentric astronomy. Copernicus finished his epoch-making work *De revolutionibus orbium coelestium* (*On the Revolutions of the Heavenly Spheres*) shortly before his death, the published copy being presented to him on his deathbed. This is the book that sparked the Scientific Revolution.

In Kepler's planetary laws and Galileo's law of falling bodies we have examples of precise mathematical laws that apply to specific systems or to a narrow range of phenomena. Today we would call these "empirical relationships" or "phenomenological laws." The genius of Newton enabled him to see behind these relationships the operation of laws of much greater generality and depth—namely, the law of universal gravitation and the three universal laws of motion.

Like all great advances in science, the articulation of these laws led to profound unifications. The first unification was of terrestrial and celestial phenomena. Until Newton, the general and deeply ingrained belief was that the heavens and the earth were wholly disparate realms governed by fundamentally different principles and even composed of different kinds of matter. The "crystalline" heavens appeared eternal, untouched by the kinds of change ("generation and corruption") that characterized the "sublunary" world. It was therefore seen as fitting that the "natural motion" of heavenly bodies should be in perfect circles, for such motion is without beginning or end (as we have seen, even Galileo could not completely free himself from these ancient ideas). What Newton showed, however, is that the very same forces govern both the celestial and the terrestrial realms. The orbits of the planets around the sun, the swinging of pendulums, and the falling of dropped weights all obey the same equations of gravity and mechanics. Indeed, Newton showed that

---

BRAHE, TYCHO (1546–1601) was born of a Danish noble family. While studying at the University of Copenhagen, his interest in astronomy was piqued by a predicted eclipse that took place in 1560. Studying existing astronomical charts, he found them all to be in disagreement with one another. At age seventeen he decided that "what is needed is a long-term project with the aim of mapping the heavens conducted from a single location over a period of several years," an effort to which (along with alchemy) he devoted his life.

In 1572, Tycho observed the appearance of a "new star" (a supernova). He was able to show that it lay far beyond the atmosphere, contradicting the Aristotelian principle that the celestial realm was unchanging. This impressed the king, who built him an observatory that Tycho named "Uraniborg" (castle of the heavens). Later, Tycho had another, subterranean observatory built nearby named "Stjeleborg" (castle of the stars).

Tycho's were the most accurate astronomical observations ever made (or that *can* be made) with the naked eye. He rejected the heliocentric Copernican theory because he understood that the motion of the earth around the sun would lead to small shifts in the apparent positions of the stars in the sky ("stellar parallax"), and he was unable to observe this. (The stars are so distant that the parallax effect was not seen until 1838.) Thus, he proposed his own geocentric model. He was aided in his last years by Kepler, who succeeded him as "Imperial Mathematicus." The vast wealth of precise data Kepler inherited from Tycho allowed him to discover that the orbit of Mars was an ellipse, not a circle, and to formulate his three great laws of planetary motion.

Tycho was an exotic figure. While a student, he lost part of his nose in a duel and wore a prosthesis made of gold and silver for the rest of his life. At his ancestral castle in Knudstrup, where he entertained on a grand scale, he kept a court jester named Jepp, a dwarf to whom Tycho attributed clairvoyance. Tycho also kept a tame moose, who, after imbibing too much beer one night at dinner, tumbled down a flight of stairs to an ignoble death.

ocean tides could be explained by the gravitational forces exerted by the moon and sun. Interestingly, this had been suggested much earlier by Kepler, but had been ridiculed as superstitious by Galileo.

MATHEMATICS IN A NEW ROLE

The Scientific Revolution, as well as the modern science to which it gave birth, was characterized not only by a happy marriage of mathematics and experiment but also a different way of looking at mathematics and its application to the physical world. A common view at the time of Copernicus and Galileo was that mathematics is *useful* for describing the quantitative aspects of things, but not particularly relevant to *understanding* those things or their causes. For example, the techniques of geometry could be used to predict accurately where heavenly bodies would appear in the sky at particular times, much as a modern train schedule is useful for predicting when trains will arrive at various stations. But just as a train schedule does not tell you what makes the trains go or why, the view of many Aristotelians was that mathematics gave no real insight into phenomena and their underlying physical causes: that was the job of "natural philosophy." (It is significant that scientists were called "philosophers" in Galileo's time, but astronomers were called "mathematicians.")

That is one reason that the heliocentric system of Copernicus did not create much of a stir before the time of Galileo. It was generally seen as merely an alternative method of computation which, while having certain advantages, involved no claim that the earth was *really* in motion. The motion of the earth in the Copernican system was widely understood to be merely "hypothetical," with no more reality than the constructions geometers made to prove their theorems. As long as any computational scheme correctly predicted where things would appear in the sky—"saved the appearances," as they put it—it was regarded as no better and no worse than any other scheme, except in terms of convenience.

Now, as a matter of fact, some of Galileo's telescopic discoveries (in particular, the phases of Venus) showed that the Ptolemaic system was no longer able even to "save the appearances," whereas the Copernican system could. That was not enough, however, to prove the earth really moved, for there was on the market an alternative, proposed by Tycho Brahe, to the Copernican and Ptolemaic systems. The system of Tycho saved all the appearances just as well as that of Copernicus, but without having to suppose that the earth moves. (It was therefore embraced by the Jesuit astronomers of the time). In fact, Tycho's system was simply the Copernican system *as viewed from the earth* (or as we might say now, it was the Copernican system as it appears in the "frame of reference" in which the earth is at rest). From a purely "mathematical" point of view, there was no way to decide between the Copernican and Tychonic systems, *and there still isn't,* if one understands the role of mathematics as most did in Galileo's time—that is, if one divorces mathematics from physical causes.

All that changed with Newton, for his laws of motion and his law of gravity provided the critical link between the *mathematical* description of motion and the *physical* causes of motion. Specifically, it related acceleration to force.

To appreciate this point, it may be helpful to consider a simple example. Suppose that a two-hundred-pound man is swinging a little ball around himself in a circle at the

---

KEPLER, JOHANNES (1571–1630) was born in Weil der Stadt, Germany. After entering the University of Tübingen to study for the Lutheran ministry, he learned of the ideas of Copernicus and became enchanted with astronomy. He taught mathematics in Graz, but was driven out by the advancing Catholic Counter-Reformation. Finding work with Tycho in Prague, he was eventually forced to leave there for the same reason. (Kepler suffered vexations from his fellow Lutherans, too. They excommunicated him for his views on the Eucharist, and, on another occasion, prosecuted his mother for witchcraft.) Using Tycho's data, Kepler discovered his three great laws of planetary motion. He was enabled to make these discoveries not only by his persistence and mathematical skill, but also by his sound physical intuition, which told him that the sun, as the largest body in the planetary system, somehow exercised a controlling influence on the other bodies. His astronomical thought was deeply influenced by his Pythagorean mysticism as well as Christian theology, which led him to see analogies between the Trinity and the interrelations of the heavenly bodies. It was Kepler who opened the door to the new science, though Copernicus and Tycho had given him the keys. At the end of his book *Harmonices mundi* (*The Harmonies of the World*), in which he announced his third law, Kepler exulted "I thank thee, Lord God our Creator, that thou allowest me to see the beauty in thy work of creation."

end of a long, elastic string. Mathematically, one can just as well consider the ball to be at rest and the man to be in circular motion around the ball. Why is the first description *physically* more sensible (as, indeed, it obviously is)? The reason is that, in the first description, we understand the *forces* that are at work, i.e., the "dynamics." Circular motion involves acceleration, and given the speed of the ball and the radius of its path we can calculate its acceleration. We can also calculate the force exerted on the ball by the string, because we understand strings. (Specifically, we can measure how much the string is stretched, and there is an empirical law, called Hooke's law, that relates the amount a string stretches to the force it exerts.) And what we would find in the sensible frame in which the ball is going round the man is that the force of the string just matches the acceleration of the ball times its mass, satisfying Newton's famous law, $F=ma$.

However, in the second description—where the ball is considered to be at rest, with the man going round the ball—the forces cannot be explained in a physically sensible way. Because the man's mass is so large, the string's force is not enough to constrain him to move in the circle. Therefore, to satisfy Newton's law in the frame of reference where the ball is at rest, large additional forces acting on the man must be assumed. Where do they come from? Nowhere. There is no intelligible physical origin for them; they must be introduced ad hoc. In modern terminology, they are "fictitious forces," and the need to postulate them is a symptom that one is not describing the situation in a *physically* sensible (or "inertial") frame of reference.

Note, then, that given the correct "dynamical laws," one may begin to understand physical reality and physical causes *through* a mathematical analysis. It is only by measuring the stretching of the string, its "elastic coefficient," the mass and speed of the ball, and the radius of its path, and then applying the dynamical laws of Newton to these quantities by means of the requisite *mathematical* calculations, that one arrives at a proper understanding of what is *physically* happening and why. (In the same way, the knowledge of Newton's law of gravity and Newton's laws of motion allows one to see that it really is the earth that is in motion about the much more massive sun.)

This is a mathematization of science far more profound than was understood before Newton, except by a few who had inklings of it, such as Kepler, Galileo, and perhaps Copernicus. When Galileo said that "the great Book of Nature is written in the language of mathematics," he heralded a radically new approach to the physical sciences.

NEWTONIAN PHYSICS

NEWTON'S LAW, $F=ma$, refers to a single point-like body, which has mass "m" and acceleration "a", and which suffers a force "F". However, it can be applied also to "extended objects" and "continuous media" by conceiving of them as made up of many small (effectively point-like) parts. Consequently, Newtonian mechanics can be used to analyze a vast range of phenomena, including the motion of fluids, the pressure of gases, the flow of heat, the vibrations of sound, and the stresses and strains of elastic solids. As these applications were made during the two centuries after Newton's laws were formulated, an ever greater unification of physics was achieved.

The universal scope and vast explanatory power of Newton's laws, as well as the technique of analyzing things into component parts, had the effect of promoting a "mechanistic" view of the world. As we have seen, the idea that the universe is a machine was commonplace even in medieval thought. However, the medieval thinkers had in mind primarily the clock-like motions of the heavenly bodies. What began to take hold in the 1600s was the idea that *everything* in nature, including plants, animals, and the human body itself, could be understood in mechanical terms.

An important aspect of this mechanistic view was the idea of "determinism." The idea of determinism in physics did not arise from anyone's philosophical prejudices or presuppositions, it came from the mathematics of Newtonian mechanics itself. In Newtonian mechanics, the state of a physical system at a particular instant of time can be completely characterized by a set of numbers, which include both "coordinates" and "momenta." The coordinates specify the positions of the parts of the system at that instant, and the momenta specify their instantaneous velocities. Given all this information at one time (the "initial conditions"), the so-called equations of motion allow one to calculate how these coordinates and momenta will evolve in time. And, generally speaking, this evolution is *unique*. This is different from, say, a game of chess, where from a particular starting position the rules of chess allow a vast number of different games to be played out. In Newtonian mechanics, if one knows the configuration (i.e., all the coordinates and momenta) at one time, then the rest of the "game" is uniquely determined from that point forward (and also backward). That is why, in 1819, the great mathematician and physicist Pierre-Simon Laplace (1749–1827) wrote, "For an intelligence which could know all the forces by which Nature is animated, and the states at some instant of all the objects that compose it, nothing would be uncertain; and the future, as well as the past, would be present to its eyes."

Newtonian physics is also "mechanistic" in the sense of dispensing with "teleology," which played so important a role in Aristotelian science. That is, in Newtonian physics the behavior of a system can be predicted without invoking any "final cause" (any future "end" toward which it is tending, or "goal" toward which it is striving). Rather, it is enough to know the *past* state of the system and the laws of physics. This fact contributed to the idea that nature is "blind" and without "purpose." It should be noted, however, that a somewhat more teleological way of looking at Newtonian physics is possible. In the eighteenth and nineteenth centuries, primarily through the work of Pierre-Louis de Maupertuis (1698–1795), Leonhard Euler (1707–83), Jean D'Alembert (1717–83), Joseph-Louis Lagrange (1736–1813), and William Rowan Hamilton (1805–65), powerful ways were developed to reformulate Newtonian mechanics in terms of the so-called "least action principle." A similar principle for optics, called the "least

---

NEWTON, SIR ISAAC (1642–1727) was born in Woolsthorpe, England, and entered Cambridge University in 1661. His genius was soon recognized by Isaac Barrow, Lucasian Professor of Mathematics, who resigned his position in 1669 so that Newton could have it. After his graduation in 1665–66, Newton, because of an outbreak of plague in Cambridge, spent an eighteen-month period in Woolsthorpe, perhaps the most productive period ever in the life of a scientist. It was then that he came up with his theory of colors, discovered the elements of both differential and integral calculus, and had crucial insights that led to his theories of gravity and mechanics.

Because of his development of the reflecting telescope, Newton was elected in 1671 as a Fellow of the Royal Society (the highest scientific honor in England). His publication, soon afterwards, of his discovery that white light is a mixture of all colors of light led to fierce controversy, which made him slow thereafter to publish his results. Nevertheless, at the urging of the astronomer Edmund Halley, Newton published his great work *Philosophiae naturalis principia mathematica* (*The Mathematical Principles of Natural Philosophy*) in 1684. It is now generally agreed that Newton and Gottfried Wilhelm Leibniz discovered calculus independently of each other and roughly simultaneously; nevertheless, a bitter dispute soon arose over whose discovery was made first, a battle that lasted long after their deaths. Newton's *Opticks* was published in 1704.

time principle," had been formulated a century earlier by Pierre Fermat (1601–65). The least time principle said that in traveling from some initial point to some final point a beam of light will follow the path that takes the least time. To solve for the light's path using this principle, one must therefore know in advance both where the light begins and where it is going to end up. The analogous principle in mechanics says that *any* system will evolve from its initial configuration to its final configuration by following the sequence of intermediate configurations (called the "trajectory," "path," or "history") that minimizes a quantity called the "action" (usually denoted S).

This way of formulating Newtonian mechanics is mathematically equivalent to the older way of formulating it in terms of forces, in the sense that it gives exactly the same answers. However, the action-principle formulation is more beautiful, powerful, and profound. In the older formulation, one starts off with as many "equations of motion" as there are coordinates needed to specify the state of the system (and for a complex system that number can be exceedingly large). With the least action principle, however, one starts off with the single fundamental quantity, S, and the requirement that the trajectory minimize it allows one to derive all the equations of motion. One thus sees another kind of unification taking place: many laws (or equations) flow from a single dynamical "principle" involving a single fundamental quantity.

Stephen M. Barr

LAPLACE, PIERRE SIMON MARQUIS DE (1749–1827) was born in Beaumont-en-Auge, France. When only eighteen, he so impressed the mathematician Jean d'Alembert with his ability that d'Alembert soon secured for him a position as professor of mathematics at the Ecole Militaire in Paris. By producing in short order a large number of papers on difficult problems in mathematics and mathematical astronomy, Laplace earned a position in the Academy of Sciences at the age of twenty-four. His magnum opus was an immense work, *Mécanique céleste*, whose five volumes appeared between 1799 and 1825. In these books he applied Newton's laws to the enormously difficult task of understanding the motions of the solar system in detail, using sophisticated mathematical techniques developed by himself and others, especially his friend the great mathematician Lagrange. One of his greatest achievements was to prove the stability of the solar system. (Newton had believed that certain instabilities required occasional readjustments by God.) When presented with a copy of *Mécanique céleste*, Napoleon asked why God was never mentioned in it, to which Laplace famously replied: "I had no need of that hypothesis." (When Lagrange heard of this, he is reported to have exclaimed, "Ah! But it is such a beautiful hypothesis; it explains many things.") Laplace also helped to lay the foundations of the theory of probability in his work *Théorie analytique des probabilités*.

Laplace is famous philosophically for his clear formulation of the idea of physical determinism. Politically, he was a survivor, currying favor with whoever was in power. This enabled him to keep his head during the French Revolution and to become, for six weeks under Napoleon, minister of the interior. Napoleon noted in his memoirs that Laplace had been "a worse than mediocre administrator, who searched everywhere for subtleties, and brought into the affairs of government the spirit of the infinitely small."

FORCES AND FIELDS

ALTHOUGH GRAVITY IS INTRINSICALLY the weakest force of nature by far, it is the force that dominates events at astronomical scales of distance and even on terrestrial scales. The reason is that gravitational forces are always attractive, so that the gravitational forces exerted on a terrestrial object (you, for instance) by the vast number of atoms in the earth all add together. By contrast, electromagnetic forces can be attractive or repulsive, and the contributions of the negatively and positively charged particles in matter tend to cancel each other out almost exactly. Thus, electromagnetic forces do not make their presence felt in daily life in an obvious way, as gravity does, even though they actually play the central role in most of the phenomena that we can directly observe. Because electromagnetic forces are more elusive and also mathematically much more complicated than Newtonian gravity, it took a long time and the work of many scientists to unravel their secrets. The most important of these were Charles Augustin de Coulomb (1736–1806), Edward Cavendish (1731–1810), Alessandro Volta (1745–1827), André Marie Ampère (1775–1836), Hans Oersted (1775–1851), Georg Simon Ohm (1789–1854), and Michael Faraday (1791–1867). The discoveries of these men were ordered, extended, and developed into a unified and coherent theory by James Clerk Maxwell (1831–79), probably the greatest physicist of the nineteenth century.

One of the crucial steps on the way to Maxwell's theory was the idea of "fields." (Interestingly, this enormously important theoretical concept was proposed not by a theorist, but by one of history's great experimentalists, Faraday.) Newton's theory of gravity was based on the idea of "action at a distance." That is, one body exerted a gravitational force directly upon another across the intervening space without any intermediary. Faraday, by contrast, conceived of there being electric and magnetic force fields filling all of space. Electric charges and electric currents produce these fields and also are acted upon by them. One can think of electric fields as being made up of "lines of force" that stretch from positive charges to negative ones and pull them together in a manner not unlike elastic bands.

As it turns out, these fields have lives of their own. They contain energy and act not only upon electrically charged particles of matter but upon each other as well. Indeed, these fields are just as real as material particles. When Maxwell completed his theory, he discovered that its equations implied that waves can propagate in these fields, and that these waves travel at the same speed as light. Indeed, subsequent experiments showed that light actually consists of such electromagnetic waves. Thus, Maxwell's theory achieved a unification of three realms of phenomena that for a long time had been thought to be quite distinct: electricity, magnetism, and optics. In fact, a far larger unification is involved, because electromagnetic forces are responsible for the interactions among and within atoms. Thus, the physical properties of matter (such as heat conductivity, elasticity, opacity, viscosity, and so on), which are studied in "Condensed Matter Physics," as well

———————

LAVOISIER, ANTOINE (1743–94), "the Father of Chemistry," was born in Paris to a wealthy family. After studying law, his interests turned to science. Although he performed many important experiments, his greatest contribution was to bring order to the theoretical chaos of chemistry. Chemists at that time labored under an upside-down theory according to which substances were thought to burn by releasing something into the air called "phlogiston," rather than by combining with something in the air, namely oxygen. Joseph Priestley, who discovered oxygen, thought it was "dephlogisticated air"; and Henry Cavendish, who discovered hydrogen, thought it was water with extra phlogiston, while oxygen was water lacking phlogiston. Lavoisier showed that combustion was really a process of oxidation, and that the whole idea of phlogiston was mistaken. At a time when many chemists still believed in four fundamental elements—air, earth, fire, and water—Lavoisier made a remarkably accurate table of thirty-three elements (only three of which turned out later to be compounds). He brought rational order also to chemical terminology, which hitherto had been totally confused. (Zinc oxide was called "flowers of zinc"; iron oxide was "astringent Mars saffron"; lead oxide was "red lead" in England and "minium" in France; sulfuric acid was "oil of vitriol," etc.) Lavoisier clarified the distinctions between, and relations among, salts, acids, oxides, and so on, and invented the modern system of chemical nomenclature. During the totalitarian madness of the Terror, charges were trumped up against Lavoisier and he was guillotined. Lagrange observed, "It required only a moment to sever that head, and perhaps a century will not suffice to produce another like it."

as the chemical properties of matter, are all based upon the electromagnetic interactions of particles.

While Maxwell's theory involved new forces, phenomena, and concepts, it is at heart a Newtonian theory. Like Newtonian mechanics it is based on a set of equations (indeed, like Newton's, they are "second-order differential equations") that deterministically govern the evolution in time of a set of coordinates and momenta. True, the notion of a coordinate must be broadened to include not only the positions of particles (as considered by Newton) but also the magnitudes of the fields that exist at every place in space. Still, Maxwell's theory involved an extension, not an abandonment, of Newtonian concepts and principles.

THE TWENTIETH-CENTURY REVOLUTIONS IN PHYSICS

THE THEORY OF RELATIVITY

The idea of the "relativity of motion" long predates Einstein—in fact, it goes back to Newtonian physics. It is the idea that velocity is not the property of one thing, but a relationship between two things. That is, it is meaningless to ask what the velocity of something is; one must ask what its velocity is *with respect to something else.* (A useful analogy is that one cannot meaningfully ask about the angle one line makes, but only about the angle between two lines.) In other words, in Newtonian physics velocity is a "relative" concept. This may seem to contradict what was said earlier about Newton's laws of motion allowing one to determine, by means of an analysis of forces, whether a ball is going around a man or the man around the ball. However, there is no real contradiction: the analysis of forces does not reveal which objects are *moving*, but rather which objects

---

MAXWELL, JAMES CLERK (1831–79) was born in Edinburgh. He graduated from Trinity College, Cambridge, in 1854, held professorships at Marischal College in Aberdeen and King's College in London, and in 1871 was named the Cavendish Professor of Physics at Cambridge University. While at Aberdeen he wrote a sixty-eight-page prize-winning paper on the nature of Saturn's rings, demonstrating that they could be stable only if made up of disconnected particles. The Astronomer Royal, Sir George Airy, described it as one of the most remarkable applications of mathematics he had ever seen. This led Maxwell to think about the motions of molecules in gases and to apply statistical methods to understanding them. He discovered, independently of Boltzmann, the "Maxwell-Boltzmann distribution" of the velocities of gas particles, and he made other fundamental contributions to statistical mechanics and thermodynamics. Inspired by Faraday's ideas, he next began to work on electricity and magnetism. He developed a theory of the dynamics of "fields" or "lines of force" that reached mathematical completion in four differential equations now called Maxwell's equations, the greatest achievement of nineteenth-century physics. Maxwell was a deeply pious man whose personality was marked by gentleness and modesty; at the same time, he was called "the most genial and amusing of companions." In his last years he selflessly nursed his ailing wife until he was rendered incapable of doing so by the cancer that killed him at the age of forty-eight.

are *accelerating*; and acceleration is an absolute concept in Newtonian physics. (It makes sense to talk about the acceleration of a single object, since acceleration is the velocity of an object at one instant *with respect to itself at another instant*.)

If one is on a train and it suddenly starts to accelerate, one can tell, even if one's eyes are closed, because one feels a force or jolt. However, if the train is not accelerating, one feels no such force, and one cannot tell whether the train is standing still or gliding perfectly smoothly with constant velocity. In fact, it is meaningless to ask whether the train is "really" moving in and of itself; one can only ask whether it is moving with respect to the platform or some other object. In the same way, if there are many objects moving uniformly with respect to one another (none of them accelerating), it makes no sense in Newtonian physics to ask which ones are really moving. What one does in practice is to choose an object on the basis of convenience and measure all motion with respect to it. This is what we meant before by picking a "frame of reference." If one is sitting in the train, it is convenient to measure motion with respect to the train; if one is standing on the platform, it is convenient to measure motion with respect to the platform. However, the "principle of relativity" says that fundamentally it does not matter what frame of reference is chosen.

What does it mean to say "it does not matter"? It certainly matters in terms of how things appear to move. To the man on the train the platform is moving, while to the man on the platform the train is moving. To put it more technically, the coordinates and momenta of objects will be different in different frames of reference. And in Newtonian physics there is a precise rule, called the "Galilean transformation law," that tells one exactly how they are different. (This rule corresponds to our everyday experiences and intuition. For example, if a car goes by you at 40 miles per hour, and another goes by you in the same direction at 75 miles per hour, then to someone in the first car, the second car will appear to be going 35—75 minus 40—miles per hour. That is what the Galilean transformation law says, and it seems to be just common sense.)

Instead, the statement "it does not matter what frame of reference is chosen" means that whatever frame one uses to measure coordinates and momenta, *the coordinates and momenta will obey the same equations*. In other words, the motion of particular objects will look different in different frames, but *the laws of physics will have the same mathematical form in any frame*. That is the key point, and the real essence, of the principle of relativity.

Maxwell's theory of electromagnetism seemed to violate the principle of relativity. It looked as though Maxwell's equations were only true if coordinates and momenta were measured in one special frame of reference. And that would appear to give a natural definition of "absolute velocity," namely velocity *as measured in that special frame*. It seemed, therefore, that there had to be something wrong with either the hallowed principle of relativity or Maxwell's theory of electromagnetism.

This is where Einstein entered the picture in 1905. His aims were actually very conservative: he did not want to abandon either the principle of relativity or Maxwell's theory. And this forced him to take a bold step. He suggested that the Galilean transformation law—the one that seems so commonsensical—is wrong, and that the correct transformation law is the one that had been formulated by a Dutch physicist named Hendrik Lorentz.

If coordinates and momenta in different frames of reference are related by the Lorentz transformation law, then it turns out that Maxwell's equations work in any frame of reference. Thus, the principle of relativity and Maxwell's theory can be reconciled. However, there is a major catch: *Newton's* laws no longer work in every frame! In other words, Einstein succeeded in saving Maxwell, but at the expense of Newton. Consequently, Newton's laws had to be changed.

What was it about Newton's laws that had to be changed? It was not his three famous laws of motion (including F=ma), nor the laws of conservation of energy and momentum, nor the "principle of least action."

All these things remain true in Einsteinian physics. Really, only one thing had to change, and that was the *geometry of space and time.* The old Galilean transformation law is based on the ideas that three-dimensional space is Euclidean in its properties, and that time is altogether distinct from space. But those ideas turned out to be wrong. The Lorentz transformation laws said—though no one had grasped their real meaning until Einstein—that space and time together make up a four-dimensional manifold that has a different kind of geometry. In this new geometry, even the Pythagorean theorem has to be modified.

We can get some idea of how time is related to space in

EINSTEIN, ALBERT (1879–1955) was born in Ulm, Germany. Contrary to romantic myth, he excelled in mathematics and physics in school, although, due to a lack of interest, he did less well in subjects that required more memorization. He entered the Swiss Federal Polytechnic School in 1896 and received his diploma in 1901. Unable to land an academic position he began work at the Swiss Patent Office. Einstein had a deep understanding of the physics of his day and the major theoretical issues confronting it, which he had been pondering for years. This bore fruit in Einstein's "miracle year," 1905, when he published three epoch-making papers: his paper proposing the theory of special relativity; his paper on the effect called "Brownian motion" (which showed that atoms are real—something still not universally accepted at that time); and his paper on the "photoelectric effect," wherein he helped lay the foundations of quantum theory by demonstrating that light acts as a particle rather than as a wave in certain situations. From 1909 to 1914 he held professorships in Zurich and Prague. In 1914 he became a professor at the University of Berlin, where he remained until Hitler came to power, at which time he renounced his German citizenship and accepted a position at Princeton University's Institute for Advanced Study. His theory of gravity—the theory of general relativity—was published in 1916. It is one of the great monuments of the human intellect. Einstein knew that he "stood on the shoulders of giants" (as Newton had said of himself). In his study he kept portraits of three men: Newton, Faraday, and Maxwell.

Einstein's theory by first considering how the three dimensions of space are related to each other. I can choose my three basic space directions (or "axes") to be "forward," "rightward," and "up." (That is to choose a frame of reference in space. It allows me to measure something's position, by saying that it is, for instance, twenty feet in front of me, thirty to the right, and ten above my head.) However, if I turn my body a little to the left, so that I am facing in a different direction than before, the direction I used to call forward, I would now have to describe as *partly* forward and *partly* rightward. In an analogous way, in Einstein's theory the "time direction" of one frame of reference becomes, in another frame of reference, *partly* the time direction and *partly* a space direction. So time and space must be thought of as four basic directions in a single four-dimensional "space-time." In this profound sense Einstein's theory *unified* space and time.

The theory of relativity led to other unifications as well. "Energy" and "mass" turned out to be, in a sense, the same thing (which is the meaning of the famous formula $E = mc^2$). And the three-component electric field and three-component magnetic field of Maxwell's theory turned out to be facets of a single, six-component "electromagnetic" field, rather than distinct entities. In fact, what is purely an electric field in one frame of reference is partly electric and partly magnetic in another frame.

As we have already noted, Newton's theory of gravity also had to be modified. This too involved a new assumption about the structure of space and time, namely that the fabric of four-dimensional space-time is curved. It is this curving or warping that is responsible for all gravitational effects in Einstein's theory of "general relativity." This makes gravity more like electromagnetism, in that gravity is no longer understood to be based on "action at a distance," as Newton said, but on fields. (The "gravitational field" at a given location is the amount that space-time is warped there.) These gravitational fields, like Maxwell's electro-magnetic fields, have lives of their own and can have waves propagating in them. The fact that all forces are now under-stood to come from fields creates the possibility of a "unified field theory" of all forces. Einstein sought such a theory in his later years without success. However, great progress has been made on this problem in recent decades.

How "Revolutionary" Was Relativity?

In what sense were Einstein's theories of special and general relativity "revolutionary"? They certainly led to conclusions that were profoundly counterintuitive and very surprising. For instance, they showed that it is not absolutely mean-ingful to say that two events happen at the "same time": it depends on the frame of reference. However, they did not completely overthrow the physics that went before; far from it. Indeed, as we saw, Einstein was led to his theory of special relativity precisely by his effort to *maintain* Maxwell's theory of electromagnetism and the old principle of relativity at the same time. Not surprisingly, therefore, Maxwell's theory was left completely untouched. And the principle of relativity was essentially untouched as well. For example, it is true in Einsteinian physics, as in Newtonian, that velocity is a relative concept while acceleration is an absolute one. Much else in Newtonian physics was also preserved, including Newton's three laws of motion and such basic concepts as force, velocity, acceleration, momentum, mass, and energy (although some of these quantities had to be reinterpreted as vectors in four-dimensional space-time rather than in three-dimensional space).

The word *revolution* is misleading when applied to scientific theories. In a revolution, the old order is swept away. However, in most of the so-called revolutions in physics, the old ideas are not simply thrown overboard, and there is not a radical rupture with the past. A better word than *revolutions* to describe these dramatic advances in science would be *breakthroughs*. In great theoretical breakthroughs, new insights are achieved that are profound, far-reaching, and take scientific understanding to a new level, a higher viewpoint. Nevertheless, many—indeed most—of the old insights retain their validity, although in some cases they are modified or qualified by new insights. Probably the only true revolution in the history of physics was the first one, *the* Scientific Revolution of the seventeenth century. The physics that preceded that revolution, namely the physics of Aristotle, was largely set aside and replaced by something thoroughly different.

We see this in the fact that physics courses in high school, college, and graduate school do not begin with a study of

Aristotelian physics. The details of Aristotelian physics are of interest only to students of history, not to modern scientists as scientists. It is not necessary to know anything at all about Aristotelian physics to do science nowadays. By contrast, before one learns the theory of relativity (or quantum theory), it is still necessary to spend several years studying the physics of the seventeenth through nineteenth centuries. That physics is still profoundly relevant. In fact, many branches of modern physics and engineering still use only pre-relativity and pre-quantum concepts.

Moreover, and this is a crucial point, the Newtonian theory of mechanics and gravity remains as *the one and only correct "limit"* of Einsteinian physics when speeds are small compared to the speed of light, and when gravitational fields are weak. That is, the smaller speeds are and the weaker gravitational fields are, the more accurately do Einstein's answers agree with Newton's answers. The idea of an older theory being the correct "limit" of the theory that replaces it is extremely important. In such a case, the older theory is not strictly speaking right; however, it is not simply wrong either. It would be better to say that it is "right, up to a point."

An example will help make this clearer. Any map of Manhattan, if it's printed on a flat piece of paper, must be wrong, strictly speaking, because the surface of the earth is curved. However, Manhattan is small enough that the earth's sphericity has negligible effects. (It affects the angles on a map of Manhattan by less than one ten-thousandth of a degree.) Indeed, it would make no sense to worry about those effects, because they are dwarfed by other ones, such as the hilliness of Manhattan Island. Therefore, ignoring the earth's sphericity is a reasonable approximation to make if one is talking about sufficiently small areas. And statements based on it are not simply falsehoods; rather, they contain real information and give correct insights into geographical relationships. This is a critical point: an incomplete and inexact description of a situation may be sufficient to convey a completely true insight into that situation. Otherwise, we could never learn anything.

Let us take another example from a "revolution" in physics that hasn't happened yet but is widely anticipated. Both Newtonian and Einsteinian physics are based on the idea that space (or space-time) is a continuous manifold of "points" that lie at definite "distances" from each other. However, quantum theory makes it appear extremely doubtful that one can apply such concepts to the very small. Many physicists expect that our intuitive concepts of space and time will prove to be altogether inadequate for describing anything smaller than a fundamental scale called the "Planck length" (about 10^{-33} cm), and that radically new concepts will have to be used. That is, a new theory will be needed. And it is thought that when this new theory is found, it will show that our intuitive concepts of "space," "time," "point," and "distance" are never applicable to the physical universe except in an approximate sense. That approximation is surely extremely good for distances much greater than the Planck length; nevertheless, it is *always* only an approximation. Supposing all these expectations someday prove to be correct, would it mean that all statements employing the concept of distance (e.g., "the book you are holding in your hands is 8 inches by 5.2 inches by 0.3 inches," or "the distance from my home to my office is 1.75 miles") are false? Obviously not. The concept of distance, while strictly speaking only approximately valid, is such a good approximation in such situations that to cavil at its use would be pedantic and unreasonable. In the same way, one is quite justified in continuing to use Newtonian physics in many situations (including all those that arise in everyday life), even though we know it does not give us an exact and complete account of what is going on.

One final point requires emphasis: Einstein's theory of relativity has nothing whatsoever to do with the foolish idea that "everything is relative." In both Newtonian physics and Einsteinian physics (as in life generally) some things are relative and some things are absolute. For instance, in both Einsteinian physics and Newtonian physics, velocity is relative but acceleration is absolute. In Newtonian physics both temporal distance and spatial distance are absolute, whereas in Einsteinian physics they are both relative, but something called "space-time distance" is absolute. And in Newtonian physics the speed of light in a vacuum is relative, whereas in Einsteinian physics it is absolute (it is the same in every reference frame). The term *relativity* has caused endless mischief. Things are not *more* relative in relativity theory than in Newtonian physics; rather, *different* things are relative and *different* things are absolute.

THE QUANTUM REVOLUTION

Quantum theory was not the brainchild of one man, as was relativity. Many great scientists contributed to its development from 1900 to the mid-1920s, when its basic structure was complete. The major founders of quantum theory include Max Planck (1858–1947), Einstein (1879–1955), Louis de Broglie (1892–1987), Niels Bohr (1885–1962), Arnold Sommerfeld (1868–1951), Max Born (1882–1970), Werner Heisenberg (1901–76), Erwin Schrödinger (1887–1961), Wolfgang Pauli (1900–1958), and Paul Dirac (1902–84).

Quantum theory has a far better claim to be considered "revolutionary" than does relativity theory. Whereas relativity changed our understanding of space and time, quantum theory fundamentally transformed the basic conceptual framework of all of physics. Physical theories that employ the pre-quantum conceptual framework (whether Newtonian or relativistic) are called "classical" theories.

Even so, it would be misleading to say that quantum theory simply "overthrew" classical physics. Quantum physics is built upon the foundations of classical physics in a profound way. In fact, there is a precise and general procedure for "quantizing" any classical theory, that is, for constructing a quantum version of it. And in the appropriate limit (roughly speaking, when systems are large) the quantum version gives the same answers as the classical version. Moreover, it is not possible to relate quantum theoretical predictions to actual measurements without making use of classical concepts.

An important fact about quantum theory is that it is based on probabilities in a fundamental way. Probabilities are often useful in classical physics, too, but there they are an accommodation to practical limitations. In classical physics, if one had complete information about a system at one time, one could (in principle) know everything about its past and future development exactly, as Laplace noted. There would be no need of probabilities. However, in quantum theory complete information about a system does *not* uniquely determine its future behavior—only the probabilities of various outcomes. This famous non-determinism (or "indeterminacy") of quantum theory is obviously of great importance philosophically. Some have argued that it is relevant in some way to the freedom of the human will, an argument that, not surprisingly, is highly controversial.

The probabilistic character of quantum theory leads to very difficult epistemological and ontological questions, which have given rise to a variety of "interpretations." The issues are too complex and subtle to review here. However, it may be of great significance that the traditional interpretation (also called the "Copenhagen," "standard," or "orthodox" interpretation) gives special status to the mind of the "observer" (i.e., the one who knows the outcomes of experiments or observations). The reason for this, in a nutshell, is that *probabilities* have to do with someone's degree of *knowledge* or lack thereof. (If one knows a future outcome, one need not use probabilities to discuss it.) As the eminent physicist Sir Rudolf Peierls (1907–95) put it, "The quantum mechanical description is in terms of knowledge, and knowledge requires *somebody* who knows." Peierls and others, such as the Nobel laureate Eugene Wigner (1902–95), have argued that the traditional interpretation of quantum theory implies that the mind of the observer

---

HEISENBERG, WERNER (1901–76) was born in Würzburg, Germany. After obtaining his doctorate in physics from the University of Munich in 1923, he worked with Max Born at the University of Göttingen and Niels Bohr at the University of Copenhagen in the rapidly developing area of quantum physics. At that point, fundamental physics was in disarray, as theorists struggled to find a consistent framework for quantum ideas to replace the existing confused patchwork of insights and methods. In 1925, at the age of twenty-three, Heisenberg discovered this framework and published his "matrix mechanics." (In 1926, Erwin Schrödinger published a "wave mechanics" that was soon shown to be the same theory in different mathematical guise. Heisenberg and Schrödinger both received the Nobel Prize in Physics.) In 1927, Heisenberg formulated his celebrated and very fundamental "uncertainty principle," which holds that the "coordinates" and "momenta" that classical physics uses to describe the state of a system cannot have definite values at the same time. He continued to make important contributions to nuclear physics, condensed matter physics, and particle physics. During World War II, he led Germany's atomic bomb project. His motives and his commitment to the project have remained the subject of controversy. However, he certainly deplored what he called the "infamies" of the Nazi regime, which he saw as a "flight into insanity that took the form of a political movement."

cannot be completely described in physical terms. If true, this assertion has profound philosophical—in particular, antimaterialist—implications. However, dissatisfaction with the traditional interpretation has led many to embrace alternatives, such as the "many worlds interpretation" or a version of the "hidden variable" or "pilot wave" theories. There is no majority view on these questions among either physicists or philosophers.

None of this philosophical confusion means that quantum theory is in any trouble as a theory of physics. There is no ambiguity or controversy about its testable predictions; and these predictions have been confirmed in countless ways over a period of eighty years, as of this writing. If superstring theory proves to be the ultimate theory of physics, as many leading physicists expect, then quantum theory is probably secure, for superstring theory does not at this point seem to entail any revision of the fundamental postulates of quantum theory.

We have observed that most great advances in physics lead to profound unifications in our understanding of nature. Quantum theory is no exception; it led to one of the most remarkable unifications of all, namely of *matter* and *forces*. In the classical electromagnetic theory of Maxwell, light is made up of waves in a field. However, Planck in 1900 and Einstein in 1905 showed that certain phenomena could not be understood unless light was assumed to come in discrete chunks, or "quanta," of energy—in other words, particles. These particles of light are now called "photons." The puzzle that something could be both a wave and a particle, a seeming contradiction, was resolved in quantum theory. "Wave-particle duality" was then found to apply across the board. Just as things that were understood classically to be waves were seen to be also particles, so things that were understood classically to be particles were now seen also to be waves—indeed, waves in a field. For instance, the electron is both a particle and a wave in an "electron field" that fills all of space. On the other hand, as Faraday taught us, forces also arise from fields. Thus, both particles of *matter* and the *forces* by which they interact are manifestations of one kind of thing, a field, which is why the basic language of fundamental physics for the last half-century is called quantum field theory. The force between two particles can be understood as being due to "field lines" stretching between them, as Faraday pictured it, or, equivalently, as being due to the exchange of "virtual particles" between them, as Richard Feynman (1918–88) pictured it.

THE ROLE OF SYMMETRY

A CRUCIAL ROLE IS PLAYED in modern physics by the idea of symmetry. In mathematics and physics the word "symmetry" has a precise definition: if a transformation of some object leaves it looking the same as before, then that transformation is said to be "a symmetry" of the object. For instance, rotating a snowflake by an angle of 60° leaves it unchanged, so one says that rotation-by-60° is a symmetry of the snowflake. Altogether, a snowflake has six such "rotational symmetries," since rotating it by 60°, 120°, 180°, 240°,

---

FARADAY, MICHAEL (1791–1867) was born in London. He was the son of a blacksmith and his education was, in his own words, "of the most ordinary description, consisting of little more than the rudiments of reading, writing, and arithmetic at a common day school." At thirteen he became an errand boy in a bookshop, and at fourteen he started a seven-year apprenticeship as a bookbinder, which gave him the opportunity to read many scientific books. In 1813, Faraday attended a public lecture by the famous chemist Sir Humphrey Davy, taking copious notes (Davy had pioneered the use of electricity to break apart compounds and had discovered in that way five chemical elements). Faraday applied for a job with him and was at first rebuffed. He soon applied again, sending Davy the notes he had taken. Impressed, Davy hired him as a secretary, then fired him (advising him to go back to bookbinding), then hired him again as a laboratory assistant. Faraday soon became a brilliant experimental chemist in his own right; however, he is most famous for his research in electromagnetism. In particular, he showed that wires moving relative to magnets had electrical currents "induced" in them. (Faraday showed that the same current was induced whether the magnet moved or the wire. This fact was one of the clues that led Einstein to his theory of relativity.) Faraday's law of induction is one of the pillars of Maxwell's theory of electromagnetism. It is even more important that Faraday was the first to articulate the concept of a force "field," an idea fundamental to modern physics. Devout, humble, and generous, Faraday is one of the most appealing personalities among the great scientists.

300°, or 360° leaves it unchanged. To take another example, a soccer-ball pattern has sixty rotational symmetries—i.e., there are sixty ways to rotate the ball that leave it looking the same. A highly developed branch of mathematics called group theory is devoted to the study of symmetry.

Symmetry is found in many of the most beautiful objects of the natural world, including flowers, shells, and crystals. It is also ubiquitous in art, architecture, music, dance, and poetry. Symmetric patterns are found in the rose windows of cathedrals, in colonnades, in friezes, in tile patterns, in arabesques, in French gardens, in the steps of dances, in the arrangements of the dancers themselves, and in the rhyme and metrical schemes of poems, to give but a few of many possible examples. Symmetry has aesthetic power because it contributes to the harmony, balance, and proportion of a thing, and also because it is a *principle of unity*. All the parts of a pattern have to be present in order for symmetry to be realized. Remove one petal of the flower, one point of the snowflake, one column in a colonnade, one rhyme in the sonnet, and the symmetry is spoiled, the unity impaired. As we shall see, symmetry is also a unifying principle in physics. The ever greater unity that we see in the laws of nature is in part the consequence of the ever deeper and more impressive symmetry principles that have been uncovered by theorists.

Symmetry can be possessed not only by physical objects, but also by something as abstract as an equation. In fact, what physicists are most interested in are the symmetries possessed by the very laws of physics themselves. Many of the great advances in physics have entailed the discovery of new fundamental symmetries of these laws. For example, the entire content of the special theory of relativity is that the laws of physics remain exactly the same in mathematical form if one shifts from one "frame of reference" to another by doing a Lorentz transformation. Thus, Einstein's theory amounts to saying that the laws of nature are "Lorentz symmetric." Another example is that Maxwell's equations of electromagnetism have a subtle symmetry called "gauge symmetry," which was discovered by the great mathematician and mathematical physicist Hermann Weyl (1885–1955). Indeed, remarkably, the very existence of electromagnetic forces in nature is a consequence of the gauge symmetry of the laws of physics.

In the twentieth century two other fundamental forces were discovered in addition to gravity and electromagnetism. They are called the "weak force" (or "weak interaction") and the "strong force" (or "strong interaction"). We do not experience them in daily life, since they are only significant at distances smaller than an atom. In the 1960s and early 1970s it was realized that these subatomic forces are also based on symmetries of the "gauge" type, and indeed owe their very existence to the presence of those symmetries in the laws of nature. Gauge symmetries are mathematically quite subtle and far removed from the kinds of symmetries that we can visualize, such as those of snowflakes or flowers. For instance, the symmetry that underlies the strong force is mathematically related to rotations that take place in an abstract space of three "complex" dimensions. (Called "complex" because the three coordinates required to locate something in such a space are not ordinary numbers, but "complex numbers." A complex number is a number of the form a + ib, where "i" is the square root of -1.)

In the early 1970s, it was further realized that the three non-gravitational forces (i.e., electromagnetic, weak, and strong) could be understood as parts of a single "grand unified" force. (While this hypothesis has not yet been confirmed definitively, there is much indirect evidence in its favor.) These theories of "grand unification" are based upon even more remarkable gauge symmetries. The symmetry of the simplest such theory, for example, involves rotations in an abstract space of *five* complex dimensions. Besides gauge symmetry, other highly recondite symmetries are suspected to be important in the fundamental laws of physics. One such is called "supersymmetry," which involves in its mathematical formulation the use of so-called Grassmann numbers. These numbers have a peculiar property: if A and B are any two Grassmann numbers, then A × B = − B × A, rather than A × B = B × A, as is the case for ordinary numbers.

As a well-known particle physicist has written,

[S]ymmetries have played an increasingly central role in our understanding of the physical world. From rotational symmetry, physicists went on to formulate ever more abstruse symmetries.... Fundamental physicists are sustained by the faith that the ultimate design is suffused with symmetries.

Contemporary physics would not have been possible without symmetries to guide us.... Learning from Einstein, physicists impose symmetry and see that a unified conception of the physical world may be possible. They hear symmetries whispered in their ears. As physics moves further away from everyday experience and closer to the mind of the Ultimate Designer, our minds are trained away from their familiar moorings....

The point to appreciate is that contemporary theories, such as grand unification or superstring, have such rich and intricate mathematical structures that physicists must marshal the full force of symmetry to construct them. They cannot be dreamed up out of the blue. Nor can they be constructed by laboriously fitting one experimental fact after another. These theories are dictated by Symmetry.[4]

"THE UNREASONABLE EFFECTIVENESS OF MATHEMATICS"

As science has progressed, the laws of nature have been found to form a unified structure, indeed a magnificent edifice of great subtlety, harmony, and beauty. As this structure is understood more deeply, it is found that ever more abstruse and yet elegant mathematics is required to describe it. Remarkably, much of that mathematics was studied and developed by pure mathematicians solely for its intrinsic interest and beauty long before it was found to be applicable to the natural world. The theory of complex numbers, for instance, was highly developed by the early 1800s even though it had no apparent relevance to science. Yet complex numbers turned out to be necessary to the formulation of quantum theory in the 1920s. Similarly, group theory was developed in the late 1800s and early 1900s, years before it was found to be useful in, and indeed of central importance to, fundamental physics. Many other striking examples could be given.

This history led Eugene Wigner to wonder about what he famously termed "the unreasonable effectiveness of mathematics" in understanding the physical world.[5] There is some deep mystery here. It seems as though the realm of pure mathematics is not something that human beings arbitrarily construct or invent, but a place in which they make discoveries. And many of the most beautiful ideas that they have discovered have proved to be exemplars or patterns for things at the most basic levels of physical reality. The more deeply we look into the heart of nature, the more mathematical it is found to be.

Since 1984, fundamental physicists have become fascinated by a theory of unprecedented mathematical depth called "superstring theory," which many of them suspect may be the long-sought unified theory of all physical phenomena. In this theory, the fundamental constituents of matter are not particles, but loops of string vibrating in a ten-dimensional space-time. Each kind of particle is, as it were, a different note played on this string. One of the greatest physicists of our time, in describing this theory to a science journalist, felt frustrated by his own inability to communicate the grandeur and magnificence of what his research had revealed to him. He said, "I don't think I've succeeded in conveying to you its wonder, incredible consistency, remarkable elegance, and beauty."[6]

The profound mathematical harmony of the world was first glimpsed by Pythagoras in his studies of the vibrations of musical strings. How appropriate it is to think that 2,500 years later the story of physics has returned to vibrating strings! The Pythagorean vision has been vindicated beyond all expectation.

Finally, consider Hermann Weyl's reflection on the mathematical elegance and beauty of science, made in a lecture at Yale University in 1931:

> Many people think that modern science is far removed from God. I find, on the contrary, that it is much more difficult today for the knowing person to approach God from history, from the spiritual side of the world, and from morals; for there we encounter the suffering and evil in the world, which it is difficult to bring into harmony with an all-merciful and all-mighty God. In this domain we have evidently not yet

succeeded in raising the veil with which our human nature covers the essence of things. But in our knowledge of physical nature we have penetrated so far that we can obtain a vision of the flawless harmony which is in conformity with sublime reason.[7]

NOTES

☙

1. Richard P. Feynman, Robert B. Leighton, and Matthew Sands, *The Feynman Lectures on Physics* (Reading, MA: Addison-Wesley, 1963), 1: 2.

2. Among those Catholic priests who made significant contributions to modern science are Christoph Scheiner (1573–1650), Niccolò Zucchi (1586–1670), Giambattista Riccioli (1598–1671), Francesco Grimaldi (1618–73), Marin Mersenne (1588–1648), Benedetto Castelli (1578–1643), Pierre Gassendi (1592–1655), Buonaventura Cavalieri (1598–1647), Niels Stensen (1638–86), Girolamo Saccheri (1667–1733), Lazzaro Spallanzani (1729–99), René-Just Haüy (1743–1822), Giuseppe Piazzi (1746–1826), Bernhard Bolzano (1781–1858), Angelo Secchi (1818–78), Gregor Mendel (1822–84), Henri Breuil (1877–1961), Julius Nieuwland (1878–1936), and Georges Lemaître (1894–1966). Information about these scientists as well as the other scientists referred to in this guide may be found in the excellent sixteen-volume *Dictionary of Scientific Biography* (New York: C. Scribner's Sons, 1970–80), as well as on many Web sites.

3. E. O. Wilson, *Consilience: The Unity of Knowledge* (New York: Knopf, 1998), 31.

4. A. Zee, *Fearful Symmetry: The Search for Beauty in Modern Physics* (New York: Macmillan, 1986), 280–81. This book is in many ways the finest introduction to the great ideas of modern physics for the layman.

5. Eugene P. Wigner, *Symmetries and Reflections: Scientific Essays of Eugene P. Wigner.* 1967. Reprint, Woodbridge, CT: Ox Bow Press, 1979.

6. John Horgan, *The End of Science: Facing the Limits of Knowledge in the Twilight of the Scientific Age* (Reading, MA: Addison-Wesley, 1996).

7. Hermann Weyl, *The Open World: Three Lectures on the Metaphysical Implications of Science* (Princeton, NJ: Princeton University Press, 1931), 28–29.

Stephen M. Barr

SUGGESTED READING

Ancient and Medieval Science

Boyer, Carl B. *A History of Mathematics*. 2nd ed. New York: Wiley, 1989.

Crombie, A. C. *The History of Science from Augustine to Galileo*. 2nd ed., 1959. Reprint, New York: Dover, 1995. (The original edition, *Augustine to Galileo*, was first published in 1952.)

———. *Robert Grosseteste and the Origins of Experimental Science, 1100–1700*. 1953. Reprint, Oxford: Oxbow, 2002.

Evans, James. *The History and Practice of Ancient Astronomy*. New York: Oxford University Press, 1998.

Grant, Edward. *The Foundations of Modern Science in the Middle Ages: Their Religious, Institutional, and Intellectual Contexts*. Cambridge: Cambridge University Press, 1996.

Lindberg, David C. *The Beginnings of Western Science: The European Scientific Tradition in Philosophical, Religious, and Institutional Context, 600 B.C. to A.D. 1450*. Chicago: University of Chicago Press, 1992.

Lindberg, David C., and Ronald L. Numbers. *God and Nature: Historical Essays on the Encounter between Christianity and Science*. Berkeley, CA: University of California Press, 1986.

The Scientific Revolution

Jardine, Lisa. *Ingenious Pursuits: Building the Scientific Revolution*. New York: Nan A. Talese, 1999.

Koestler, Arthur. *The Sleepwalkers: A History of Man's Changing Vision of the Universe*. 1959. Reprint, Harmondsworth: Penguin, 1986. (This triple biography of Copernicus, Kepler, and Galileo is somewhat slanted against Galileo, but especially good on Kepler.)

Koyré, Alexandre. *From the Closed World to the Infinite Universe*. 1957. Reprint, Baltimore: Johns Hopkins University Press, 1994.

Langford, Jerome J. *Galileo, Science, and the Church*. 1966. Reprint, South Bend, IN: St. Augustine's Press, 1998. (This is a readable and judicious account of the Galileo affair that is historically, scientifically, and theologically accurate.)

Principe, Lawrence. *The Aspiring Adept: Robert Boyle and his Alchemical Quest*. Princeton, NJ: Princeton University Press, 1998.

Sobel, Dava. *Galileo's Daughter: A Historical Memoir of Science, Faith, and Love*. New York: Walker, 1999.

Westfall, Richard S. *Never at Rest: A Biography of Isaac Newton*. 1980. Reprint, New York: Cambridge University Press, 1998. (This reprint contains the author's preface for the 1983 edition, as well as the original preface.)

ELECTROMAGNETISM

Cantor, Geoffrey. *Michael Faraday: Sandemanian and Scientist*. New York: St. Martin's Press, 1991.

Mahon, Basil. *The Man Who Changed Everything: The Life of James Clerk Maxwell*. Chichester, Eng.: Wiley, 2003.

TWENTIETH-CENTURY PHYSICS

Feynman, R. P. *"Surely You're Joking, Mr. Feynman!": Adventures of a Curious Character*. 1985. Reprint, New York: W. W. Norton, 1997. (Here is an amusing autobiography by one of the great figures of twentieth-century physics.)

Hoffmann, Banesh. *Albert Einstein, Creator and Rebel*. New York: Viking Press, 1972.

Lovett, Barbara Cline. *The Questioners: Physicists and the Quantum Theory*. 1965. Reprint, Chicago: University of Chicago Press, 1987. (This is a popular history of quantum theory and the men who developed it.)

Segrè, Emilio. *Enrico Fermi—Physicist*. Chicago: University of Chicago Press, 1970.

Zee, A. *An Old Man's Toy: Gravity at Work and Play in Einstein's Universe.* 1989. Reprint, New York: Oxford University Press, 2001.

———. *Fearful Symmetry: The Search for Beauty in Modern Physics.* 1986. Reprinted with a new preface and afterward, Princeton, NJ: Princeton University Press, 1999.

Astronomy

Grosser, Morton. *The Discovery of Neptune.* 1962. Reprint, New York: Dover, 1979.

Kolb, Edward W. *Blind Watchers of the Sky: The People and Ideas that Shaped Our View of the Universe.* Reading, MA: Addison-Wesley, 1996. (Written by a cosmologist, this wry history is especially strong on the modern era.)

Sobel, Dava. *Longitude: The True Story of a Lone Genius Who Solved the Greatest Scientific Problem of His Time.* 1995. 10th anniv. ed., New York: Walker, 2005.

Geology and Paleontology

Cutler, Alan. *The Seashell on the Mountaintop: A Story of Science, Sainthood, and the Humble Genius Who Discovered a New History of the Earth.* New York: Dutton, 2003. (This is a biography of Niels Stensen/Nicolaus Steno.)

Oreskes, Naomi. *Plate Tectonics: An Insider's History of the Modern Theory of the Earth.* Boulder, CO: Westview Press, 2001.

Rudwick, Martin J. S. *The Great Devonian Controversy: The Shaping of Scientific Knowledge among Gentlemanly Specialists.* Chicago: University of Chicago Press, 1985.

———. *The Meaning of Fossils: Episodes in the History of Palaeontology.* 1972. Reprint, Chicago: University of Chicago Press, 1985.

———. *The New Science of Geology: Studies in the Earth Sciences in the Age of Revolution.* Burlington, VT: Ashgate, 2004.

CHEMISTRY

Hoffmann, Roald, and Vivian Torrence. *Chemistry Imagined: Reflections on Science*. Washington, DC: Smithsonian Institution Press, 1993.

Jaffe, Bernard. *Crucibles: The Lives and Achievements of the Great Chemists*. 1930. Reprint, New York: Tudor, 1936.

Levere, Trevor H. *Transforming Matter: A History of Chemistry from Alchemy to the Buckyball*. Baltimore: Johns Hopkins University Press, 2001.

Morris, Richard. *The Last Sorcerers: The Path from Alchemy to the Periodic Table*. Washington, DC: Joseph Henry, 2003.

BIOLOGY

Browne, E. Janet. *Charles Darwin: A Biography.* Vol. 1, *Voyaging*. New York: Knopf, 1995.

———. *Charles Darwin: A Biography*. Vol. 2, *The Power of Place*. New York: Knopf, 2002.

de Kruif, Paul. *Microbe Hunters*. 1926. 70th anniv. ed., San Diego: Harcourt, Brace, 1996. (De Kruif's book has inspired many people to pursue careers in science.)

Gould, Stephen Jay. *Wonderful Life: The Burgess Shale and the Nature of History*. 1989. Reprint, London: Vintage, 2000.

Larson, Edward J. *Evolution: The Remarkable History of a Scientific Theory*. New York: Modern Library, 2004.

———. *Summer for the Gods: The Scopes Trial and America's Continuing Debate over Science and Religion*. New York: Basic Books, 1997. (This book disposes of many myths.)

Morris, Simon Conway. *The Crucible of Creation: The Burgess Shale and the Rise of Animals*. New York: Oxford University Press, 1998.

Stephen M. Barr

SCIENCE AND PHILOSOPHY

Barrow, John D., and Frank J. Tipler. *The Anthropic Cosmological Principle*. 1986. Reprint of the corrected 1988 ed., New York: Oxford University Press, 1996.

Davies, P. C. W., and Julian R. Brown. *The Ghost in the Atom: A Discussion of the Mysteries of Quantum Physics*. 1986. Reprint, New York: Cambridge University Press, 1999. (This book contains interviews with physicists representing the various schools of thought on "the interpretation of quantum theory.")

Heisenberg, Werner. *Across the Frontiers*. 1974. Reprint, Woodbridge, CT: Ox Bow Press, 1990. (See this book for a brilliant essay on beauty in science.)

Polanyi, Michael. *Personal Knowledge: Towards a Post-critical Philosophy*. 1958. Reprint, London: Routledge, 1998. (This is a great work on the philosophy of science and epistemology.)

Shapiro, Stewart. *Thinking about Mathematics: The Philosophy of Mathematics*. New York: Oxford University Press, 2000.

Weyl, Hermann. *The Open World: Three Lectures on the Metaphysical Implications of Science*. 1932. Reprint, Woodbridge, CT: Ox Bow Press, 1989.

Wigner, Eugene P. *Symmetries and Reflections: Scientific Essays of Eugene P. Wigner*. 1967. Reprint, Woodbridge, CT: Ox Bow Press, 1979. (Here are important essays on the role of mathematics in science, the mind-body question, and the interpretation of quantum theory.)

INTERCOLLEGIATE
STUDIES INSTITUTE
Educating for Liberty

ISI Books is the publishing imprint of the Intercollegiate Studies Institute, a nonprofit, nonpartisan organization whose mission is to inspire college students to discover, embrace, and advance the principles and virtues that make America free and prosperous.

Founded in 1953, ISI teaches future leaders the core ideas behind the free market, the American Founding, and Western civilization that are rarely taught in the classroom.

ISI is a nonprofit, nonpartisan, tax-exempt educational organization. The Institute relies on the financial support of the general public—individuals, foundations, and corporations—and receives no funding or any other aid from any level of the government.

To learn more about ISI,
visit **www.isi.org** or call **(800) 526-7022**